FIELD GUIDES
FOR KIDS

FISH

Christopher Forest

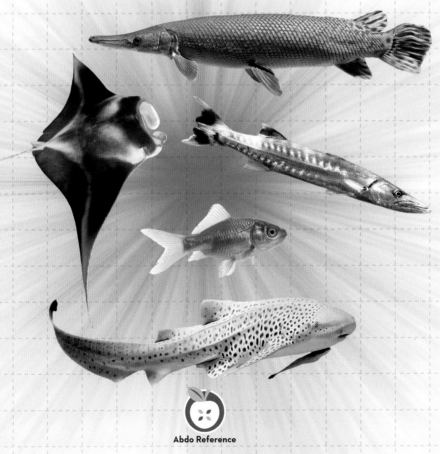

Abdo Reference

An Imprint of Abdo Publishing | abdobooks.com

CONTENTS

WHAT IS A FISH?

Fish is a general term used to identify more than 27,000 species of animals that inhabit fresh water and salt water. Fish are diverse, and they come in all shapes and sizes. For instance, the smallest fish are 0.3 inches (0.76 cm) long, while the largest fish span 39 feet (12 m).

CLASSIFYING FISH

Classifying fish into species can be difficult because they are very diverse. However, most fish have some common characteristics:

- All fish spend at least some time in water. They have heads with obvious eyes, sense organs, and teeth. Fish also have brains that are protected by braincases.

- Fish are classified as vertebrates, which means they have backbones.

- A fish's skin often contains scales that help protect it from injuries and predators.

- Many fish have fins that come in pairs. These limbs help to propel fish through the water.

- Most fish are cold-blooded. This means they cannot regulate their own body temperatures. They depend on their environments and movements to generate heat.

- Fish have gills that can be used to get oxygen from water.

- Many fish species have a sense organ called a lateral line system. This system helps fish find prey or sense predators.

- Fish may also have swim bladders that help them remain buoyant.

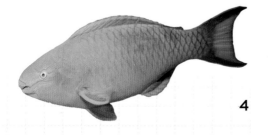

Some fish do not share all of these characteristics. For example, tuna are warm-blooded fish. Lampreys are classified as invertebrates because they lack backbones. Lungfish can use their lungs to breathe.

CATEGORIES OF FISH

Fish fall into three specific categories:

- Agnatha fish are primitive fish, which means they have existed for many millions of years. In fact, scientists believe these were the first type of fish to appear on Earth. The first Agnatha appeared 450 million years ago. Agnatha fish do not have jaws.

- Chondrichthyes are fish that are made up of cartilage. These include sharks, rays, and skates.

- Osteichthyes have bones and are the largest group of fish. They are divided into ray-finned fish and lobe-finned fish.

HOW TO USE THIS BOOK

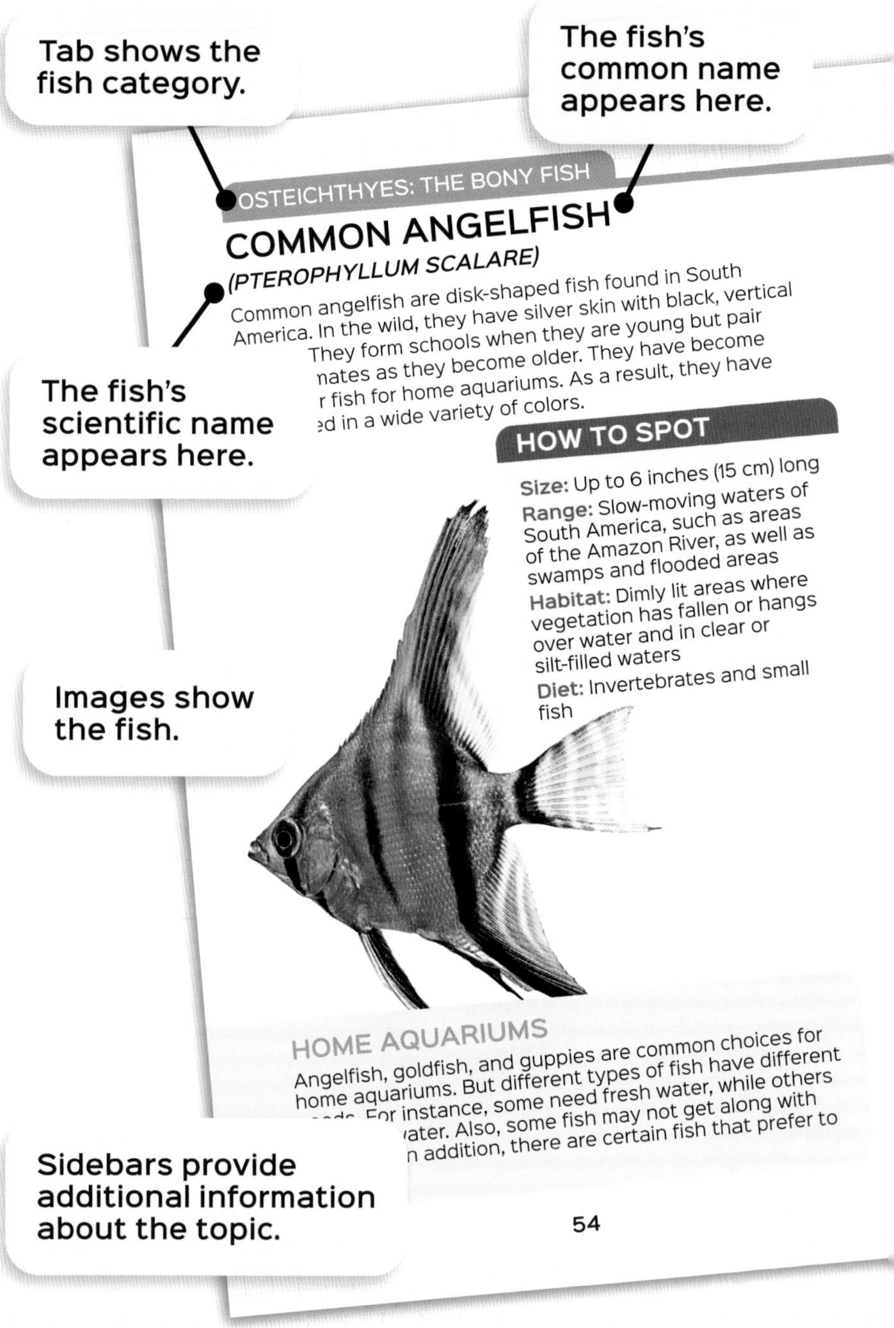

Tab shows the fish category.

The fish's common name appears here.

OSTEICHTHYES: THE BONY FISH

COMMON ANGELFISH
(PTEROPHYLLUM SCALARE)

Common angelfish are disk-shaped fish found in South America. In the wild, they have silver skin with black, vertical They form schools when they are young but pair nates as they become older. They have become r fish for home aquariums. As a result, they have ed in a wide variety of colors.

The fish's scientific name appears here.

HOW TO SPOT

Size: Up to 6 inches (15 cm) long

Range: Slow-moving waters of South America, such as areas of the Amazon River, as well as swamps and flooded areas

Habitat: Dimly lit areas where vegetation has fallen or hangs over water and in clear or silt-filled waters

Diet: Invertebrates and small fish

Images show the fish.

HOME AQUARIUMS

Angelfish, goldfish, and guppies are common choices for home aquariums. But different types of fish have different ads. For instance, some need fresh water, while others vater. Also, some fish may not get along with n addition, there are certain fish that prefer to

Sidebars provide additional information about the topic.

54

6

COMMON CARP *(CYPRINUS CARP*

The common carp is a freshwater fish originally fro
and Asia. Today, it is spread throughout North Am
have brassy-green skin with yellow underbellies. T
slimy skin, large scales, and long bodies. Each car
two barbels attached to its upper jaw.

> **This paragraph provides information about the fish.**

HOW TO SPOT

Size: 12 to 24 inches
(30 to 60 cm) long;
1 to 9 pounds (0.4 to 4

Range: Every contine
except Antarctica

Habitat: Wide variety
waters, including stre
and reservoirs

Diet: Plants and mar
life such as crayfish,
leeches, and shrimp

> ***How to Spot* features give information about the fish's size, range, habitat, and diet.**

FUN FACT
Some scientists
estimate that
a female carp
can lay around
300,000 eggs each
breeding season.

> ***Fun Facts* give interesting information related to fish.**

INVASION OF THE COMMON CARP
The common carp was originally from Europe and Asia.
But it probably came to North America in the 1800s. It was
introduced as a game fish in the United States. Now, carp
can be found in almost every US state.

PACIFIC HAGFISH
(EPTATRETUS STOUTII)

Pacific hagfish are primitive, soft, eel-like fish that feed on other fish. Most have tentacles on the ends of their snouts. They also have five to 15 pairs of gills. They have no eyes. These fish live on the bottom of the Pacific Ocean. They are known for attaching to and feeding on dead or injured fish. They burrow into the fish. Then, they eat the fish from the inside out.

HOW TO SPOT

Size: Up to 25 inches (64 cm) long; 1.8 to 3.1 pounds (0.8 to 1.4 kg)

Range: Pacific Ocean

Habitat: Cold ocean waters at depths of 4,280 feet (1,300 m); live in burrows

Diet: Ocean invertebrates and dead or dying fish

FUN FACT

A hagfish produces so much slime that it is sometimes called the slime eel. The slime is used for defense. When threatened, the hagfish will release the slime from its nostrils. It will fill the gills of a nearby predator.

SEA LAMPREY *(PETROMYZON MARINUS)*

Sea lampreys are ancient jawless, boneless fish. They reproduce in fresh water, but many eventually move to salt water. Sea lampreys tend to have one or two dorsal fins and seven gill openings on the sides of their bodies. Their mouths have horn-like teeth and sucking devices that allow them to attach to host fish like suction cups. Then, they eat the tissue, fluid, and blood of the host fish.

HOW TO SPOT

Size: Freshwater sea lampreys can get up to 25 inches (64 cm) long. Ocean sea lampreys can get up to 47 inches (120 cm) long. The largest lampreys can reach 2.25 pounds (1 kg).

Range: Western and eastern Atlantic coasts and Great Lakes region

Habitat: Fresh water or salt water

Diet: Host fish, such as trout

WHAT IS AN AGNATHA?

Scientists disagree if Agnatha fish should be classified as true fish. That's because they lack some features of modern fish species. For instance, they have circular mouths and are generally missing pairs of fins. However, they do have fins and gills, which are the standard features of fish.

Unlike many fish, Agnatha do not have internal skeletons made of bone. In fact, some species don't have skeletons at all. But for those species that do, their skeletons are made of mostly cartilage.

9

BASKING SHARK
(CETORHINUS MAXIMUS)

Basking sharks live in cold and temperate oceans. Surface water in temperate oceans is between 50 and 68 degrees Fahrenheit (10 and 20°C). These sharks are black, gray, or brownish gray. They are slow moving, and they have wide mouths that help them catch zooplankton. Basking sharks often swim in groups near the surface of the ocean. They migrate throughout the year.

HOW TO SPOT

Size: Up to 30 to 40 feet (9 to 12 m) long; 5 tons (4,500 kg)
Range: Temperate and Arctic oceans
Habitat: Near the shore and at the surface of the water
Diet: Zooplankton

FUN FACT
Basking sharks get their name because they look like they are bathing in sunlight near the surface of the ocean.

COMMON THRESHER SHARK

(ALOPIAS VULPINUS)

Common thresher sharks have bluish-gray, brown, gray, or black skin with white underbellies. They are aggressive hunters that have been known to attack boats. However, they tend to avoid humans. These sharks are known for their distinctive tails, which are shaped like sickles. They are sometimes called sea foxes or thrashers.

HOW TO SPOT

Size: 8.5 to 15 feet (2.6 to 4.5 m); up to 750 pounds (340 kg)

Range: Cold-temperate and tropical ocean waters around the world

Habitat: Open ocean waters and coastal regions

Diet: Fish, squid, and seabirds

SHARK ATTACKS

Most shark attacks happen when a shark mistakes a human for another animal. There are only between 70 and 100 shark attacks worldwide each year. The likelihood of dying in a shark attack is even slimmer. The odds are only one in more than three million.

GOBLIN SHARK *(MITSUKURINA OWSTONI)*

The goblin shark is one of the rarest sharks in the world. They have an almost ghostly appearance, with soft, pinkish-white skin and blue-gray fins. Young sharks are born white and turn pink over time. The shark has a long snout, tiny eyes, and five gill openings on either side of its body. When the fish is hunting, it projects its jaw out to catch prey.

HOW TO SPOT

Size: Up to 20 feet (6 m) long

Range: Throughout the world's oceans

Habitat: Deep in the ocean, often near continental slopes

Diet: Fish, crustaceans, octopuses, shrimp, and squid

FUN FACT

The goblin shark is the only member of the *Mitsukurinidae* family that's still in existence.

GREAT HAMMERHEAD
(SPHYRNA MOKARRAN)

The great hammerhead shark is the largest species of hammerheads. These sharks tend to be brown or gray with cream-colored underbellies. The shark's standout feature is the flat head that sticks out from its body. The head has special organs called the ampullae of Lorenzini. This allows the hammerhead to detect electric signals from animals it might prey upon. Both sharks and rays have these sense organs.

HOW TO SPOT

Size: Up to 18 to 20 feet (5.5 to 6 m) long; up to 991 pounds (450 kg)

Range: Warm-temperate and tropical waters

Habitat: Offshore in depths of up to 980 feet (300 m) and in shallow areas along coasts

Diet: Fish and stingrays

GREAT WHITE SHARK
(CARCHARODON CARCHARIAS)

Great white sharks can have gray, dark blue, or brown coloring on their sides and backs. Their bellies are white. Their torpedo-like bodies help make them the largest predator fish in the world. Great whites are dangerous hunters with sharp, triangular-shaped upper teeth. These creatures can leap out of the water to snatch their food.

FUN FACT
About one-third to one-half of all shark attacks on humans are reportedly caused by great whites. However, these sharks do not actively hunt humans.

HOW TO SPOT

Size: Up to 15 to 20 feet (4.5 to 6 m) long; 5,000 pounds (2,300 kg)
Range: Mostly in temperate oceans around the world
Habitat: Often found near shores, but also at depths greater than 4,200 feet (1,300 m)
Diet: Mostly marine mammals and fish

INSPIRATION FOR THE MOVIE JAWS

Jaws is a famous movie about a great white shark attacking people. The movie was inspired, in part, by a series of events that occurred in 1916. That year, at least one great white shark prowled the waters off of New Jersey. Over the course of two weeks, the shark—or perhaps more than one shark—attacked five people. Four of those victims died because of the attacks. One great white shark was actually caught off the New Jersey coast. Its stomach contained human remains.

LEMON SHARK
(NEGAPRION BREVIROSTRIS)

Lemon sharks are powerful, and they live in the shallow waters of the Pacific and Atlantic Oceans. They have yellowish or olive-colored skin that helps them blend in with their sandy surroundings. They have been seen swimming into some freshwater bodies. Lemon sharks have stocky bodies and round, blunt noses. A lemon shark has a special type of eye. This helps the shark see different colors and shadings.

HOW TO SPOT

Size: 8 to 10 feet (2.4 to 3 m) long; 550 pounds (250 kg)

Range: Regions of the Atlantic Ocean—from New Jersey to the Caribbean, Senegal, and Ivory Coast—and northern Pacific Ocean

Habitat: Shallow coastal waters, coral reefs, and mangrove forests

Diet: Crustaceans, fish, rays, seabirds, and sharks

MEGAMOUTH SHARK
(MEGACHASMA PELAGIOS)

Megamouth sharks are dark brown with silvery-white underbellies. The shark takes in food through its large mouth and filters it through its gills. A typical megamouth might have a mouth that can open 4 feet (1.2 m) wide or more.

FUN FACT

Megamouth sharks live in the deep seas and are believed to only come to the surface of the ocean at night. As a result, less than 70 of these sharks have been officially observed.

HOW TO SPOT

Size: 17 feet (5.2 m) long; 2,700 pounds (1,200 kg)

Range: Temperate and tropical regions of the Atlantic, Pacific, and Indian Oceans

Habitat: Can live up to 15,000 feet (4,600 m) deep

Diet: Krill

NURSE SHARK
(GINGLYMOSTOMA CIRRATUM)

Nurse sharks are large and slow moving. They live in warm, shallow waters. Their skin tends to be yellowish brown to dark brown, and young nurse sharks have spots. A nurse shark can grow several feet long, with 25 percent of its length composed of its tail fin. A nurse shark has two barbels extending from its snout. These barbels help the shark find food.

HOW TO SPOT

Size: Up to 10 feet (3 m) long; 230 pounds (100 kg)
Range: Tropical and subtropical waters off both North American coasts
Habitat: Ocean bottoms and near coral reefs
Diet: Fish, invertebrates, and rays

FUN FACT
Nurse sharks can use their fins to crawl along the ocean bottom.

SHORTFIN MAKO SHARK
(ISURUS OXYRINCHUS)

The shortfin mako shark is a large, dangerous predator. These sharks have gray or blue bodies with white underbellies. The mako has a streamlined head and body, making it ideally suited to move swiftly through the water. It also can leap out of the water, making it a fierce hunter. Mako sharks are sometimes called sharp-nosed mackerel sharks or blue pointers.

HOW TO SPOT

Size: Up to 12 feet (3.6 m) long; at least 1,200 pounds (545 kg)

Range: Tropical and temperate oceans

Habitat: Waters ranging from the surface to around 1,640 feet (500 m) deep

Diet: Variety of fish, including herring, mackerel, and swordfish

TYPES OF SHARK TEETH

Sharks have different types of teeth. Some are sharp and pointed. Others are triangular shaped with grooves on the edges. Some sharks have wide, flat teeth, and others have giant filters. The type of teeth a shark has depends on what kind of food it eats.

TIGER SHARK *(GALEOCERDO CUVIER)*

The tiger shark is one of the largest sharks in the world. It is gray, and when it is young there are dark spots or lines on its sides. Tiger sharks move slowly, but they can effectively ambush prey. However, tiger sharks usually get food by scavenging anything they can find.

HOW TO SPOT

Size: 10 to 14 feet (3 to 4.2 m) long; 850 to 1,400 pounds (385 to 635 kg)

Range: Tropical and temperate waters of the world, except for the Mediterranean Sea

Habitat: Mainly murky coastal waters, but also inlets, harbors, and lagoons

Diet: Bony fish, clams, dolphins, rays, seals, sea snakes, and sea turtles

FUN FACT

Tiger sharks have been known to eat garbage. License plates and tires have been found in the remains of dead tiger sharks.

WHALE SHARK *(RHINCODON TYPUS)*

The whale shark is the largest fish in the ocean. Whale sharks have dark brown or gray skin with pale stripes and white spots. These slow-moving sharks can grow to lengths of a typical school bus. While it is large in size, a whale shark is a filter feeder, posing little threat to humans or other fish in the ocean. Its large jaw allows it to constantly pass water through its mouth and gills to get food.

HOW TO SPOT

Size: Can range from 18 to 32 feet (5.5 to 9.8 m) long; up to 20.6 tons (18,700 kg)

Range: All the world's tropical oceans

Habitat: Feed at the surface but can dive more than 3,280 feet (1,000 m)

Diet: Plankton

FUN FACT

The spot patterns on whale sharks are like human fingerprints. No two whale sharks' patterns are exactly alike.

ZEBRA SHARK
(STEGOSTOMA FASCIATUM)

The zebra shark is a bottom-dwelling shark that lives in coastal regions. The coloring of the zebra shark changes as it grows. It begins life with dark skin and white or yellow stripes. However, as it ages its skin lightens to a yellow-brown color and becomes spotted. The shark looks torpedo shaped, with ridges running down its sides. Zebra sharks have small mouths—ideal for eating mollusks. They have barbels in the front of their mouths that help them locate food.

HOW TO SPOT

Size: Up to 8 feet (2.4 m) long; 44 to 66 pounds (20 to 30 kg)

Range: Indian and Pacific Oceans

Habitat: Coastal regions on coral bottoms, sandy depths, or rock reefs

Diet: Crustaceans, mollusks, and small fish

ATLANTIC DEVIL RAY
(MOBULA HYPOSTOMA)

Atlantic devil rays live in the coastal areas of the Atlantic Ocean. These rays have triangular-shaped pectoral fins. They often curl their fins as they move through the water. These rays have dark gray or black skin with cream-colored underbellies. Devil rays often travel in large groups, called schools. They are sometimes called devil fish, diablos, or lesser devil rays.

FUN FACT

Atlantic devil rays got their name from their curled-up fins, which can look like horns.

HOW TO SPOT

Size: Up to 4 feet (1.2 m) long

Range: Atlantic Ocean from North Carolina to Argentina

Habitat: Coastal regions, but occasionally open waters

Diet: Crustaceans and small fish

ATLANTIC STINGRAY
(DASYATIS SABINA)

The Atlantic stingray has brown or yellowish-brown skin with a lighter-colored underbelly. It is one of the smallest stingrays in the world. These rays can detect electric signals that help them find their prey. An Atlantic stingray has a long, slender tail with venomous spines. The spines are used by the ray to defend itself.

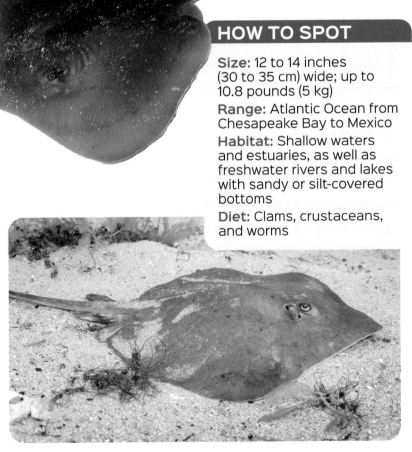

HOW TO SPOT

Size: 12 to 14 inches (30 to 35 cm) wide; up to 10.8 pounds (5 kg)

Range: Atlantic Ocean from Chesapeake Bay to Mexico

Habitat: Shallow waters and estuaries, as well as freshwater rivers and lakes with sandy or silt-covered bottoms

Diet: Clams, crustaceans, and worms

BAT RAY *(MYLIOBATIS CALIFORNICA)*

The bat ray has smooth, dark brown or black skin with a white underbelly. The ray gets its name from the pectoral fins that spread out much like a bat's wings. Bat rays often use these fins to fan the sandy ocean bottom in search of food. This ray also has a tail with a venomous spine.

HOW TO SPOT

Size: Up to 6 feet (1.8 m) wide; up to 200 pounds (91 kg)

Range: Eastern Pacific Ocean from Oregon to the Gulf of California, as well as near the Galápagos Islands

Habitat: Bottom of bay floors in sandy areas, near coral reefs, and in kelp forests

Diet: Crustaceans, mollusks, and small fish

BULLNOSE RAY
(MYLIOBATIS FREMINVILLII)

Bullnose rays are brown or gray with white underbellies. This small ray has a head that sticks out from its body and resembles the bill of a duck. It has a whip-like tail. The base of its spine is sharp. Bullnose rays occasionally leap out of the water as they swim. They are sometimes called blue-nosed rays or eagle rays.

FUN FACT
Newborn bullnose rays are only around 10 inches (25 cm) wide.

HOW TO SPOT

Size: Around 3 feet (0.9 m) wide

Range: Atlantic Ocean from Massachusetts to Florida

Habitat: Typically shallow coastal waters, such as estuaries

Diet: Mollusks and crustaceans such as crabs and lobsters

COMMON EAGLE RAY
(MYLIOBATIS AQUILA)

The common eagle ray has dark, copper-colored or black skin on its back with a white underbelly. The ray is known for its large fins, which make it much wider than it is long. While often considered relatively harmless to humans, this ray does have a sharp spine at the base of its tail that can injure people. These rays are sometimes called sea eagles and toadfish and often live in small schools.

HOW TO SPOT

Size: Up to 6 feet (1.8 m) wide; up to 32 pounds (14.5 kg)

Range: Eastern Atlantic Ocean from the British Isles to South Africa, and the Indian Ocean along the southeastern coast of Africa

Habitat: Shallow bays, estuaries, and lagoons

Diet: Crustaceans and mollusks

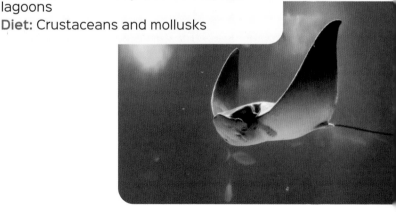

COWNOSE RAY *(RHINOPTERA BONASUS)*

The cownose ray is a kite-shaped ray with brown or olive skin and a white underbelly. This ray is large with a brown, whip-like tail. The square snout of this ray projects from its face and resembles a cow's nose. This is how the ray got its name. Cownose rays are known to leap out of the water and land on the surface, creating a loud smacking sound. They have flat teeth, which allow them to prey on soft-shelled animals. They often form schools based on gender and age.

FUN FACT

Some people mistake cownose rays for sharks. When these rays swim, their fins sometimes poke out of the water's surface and look like a shark's dorsal fin.

HOW TO SPOT

Size: Wingspan can be up to 3 feet (0.9 m) wide; up to 50 pounds (23 kg)

Range: Western Atlantic Ocean and the northwestern coast of Africa

Habitat: Muddy coastal waters and brackish waters, where the salt water of the ocean meets the fresh water of rivers

Diet: Clams and oysters

GIANT MANTA RAY *(MANTA BIROSTRIS)*

Giant manta rays are triangular shaped with dark skin on top and white underbellies. These rays also have markings on their backs and underneath their bodies. Giant manta rays get their name for their giant wingspans. In fact, these rays are one of the largest fish in the world. Manta rays are filter feeders, taking in water through their mouths and filtering out plankton for food. Some rays have been known to leap entirely out of the water to gain the attention of a mate or to play with other rays.

HOW TO SPOT

Size: Wingspans reach up to 29 feet (9 m) wide; up to 5,300 pounds (2,400 kg)

Range: Worldwide in temperate, subtropical, and tropical oceans

Habitat: Coastal regions and open oceans

Diet: Plankton and other small food

FUN FACT

Manta rays often travel in large groups. These groups are called squadrons.

LESSER ELECTRIC RAY
(NARCINE BANCROFTII)

Lesser electric rays have a circular shape with dark brown to reddish skin and pale-colored underbellies. As their name suggests, these rays have a special organ in their bodies that can produce an electric current. This current can be used to shock prey and ward off predators.

HOW TO SPOT

Size: Up to 33 inches (84 cm) long

Range: Western Atlantic coast from North Carolina to Argentina, and there have been some reported sightings in the Yucatán region of the Pacific Ocean

Habitat: Shallow coastal waters in beds of seagrass, as well as muddy or sandy ocean floors

Diet: Crustaceans, marine worms, and shrimp

ELECTRIC FISH

Many fish can both produce and detect electricity. These electric fish are often categorized by the type of electricity they use. Strong electric fish produce a strong current. These include electric catfish, electric eels, and electric rays. Weak electric fish produce a weak current that is often less than one volt. These include Peter's elephantnose fish and the knife fish. Some fish only detect electricity. These include catfish, paddlefish, sharks, skates, and most rays.

ROUND STINGRAY *(UROBATIS HALLERI)*

Round stingrays tend to have gray skin, and some have spots or are mottled. Their underbellies can be white, yellow, orange, or yellow orange. These rays have rounded bodies with slightly pointed snouts. They also have rounded tail fins, which is uncommon for most rays.

HOW TO SPOT

Size: 8 to 10 inches (20 to 25 cm) long; up to 3 pounds (1.4 kg)

Range: Eastern Pacific Ocean from California to Panama

Habitat: Regions of shallow ocean water with sandy floors

Diet: Clams, sea worms, small fish, and snails

SMOOTH BUTTERFLY RAY
(GYMNURA MICRURA)

A smooth butterfly ray has gray, green, or light brown skin with spots. This allows the ray to blend into its environment. These rays are diamond shaped, and they are wider than they are long. The coloring, spots, and shape make the rays resemble butterflies. They do not possess any spines, which limits their danger to humans. They are sometimes called diamond skates or skeetes.

FUN FACT
The smooth butterfly ray can change shades to blend in with its background.

HOW TO SPOT

Size: About 3 to 4 feet (1 to 1.2 m) wide

Range: Western Atlantic Ocean from Maryland to Brazil, the Gulf of Mexico, and the eastern Atlantic Ocean from Senegal to the mouth of the Congo River

Habitat: Coastal regions on sandy or muddy ocean floors

Diet: Invertebrates and small fish

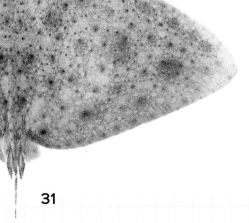

SPINY BUTTERFLY RAY
(GYMNURA ALTAVELA)

The diamond-shaped spiny butterfly ray is wider than it is long. These rays have short tails compared to other rays, with one or two barbs. They tend to be dark brown with lines, spots, patterns, or swirls. The color and shape of the ray resembles a butterfly. The coloring of the ray helps it blend in with its ocean-bottom surroundings.

HOW TO SPOT

Size: 7 feet (2 m) wide; up to 130 pounds (60 kg)

Range: Western Atlantic Ocean from Massachusetts to Argentina and the eastern Atlantic Ocean from Portugal to Angola

Habitat: Ocean floor bottoms near sand or mud, rocky reefs, and estuaries

Diet: Crustaceans, fish, mollusks, and plankton

YELLOW STINGRAY
(UROBATIS JAMAICENSIS)

A yellow stingray has a round shape with a short tail and venomous spine. These rays can be various colors that allow them to blend in with the sandy ocean bottom. Some rays have giraffe-like patterns of green or dark brown on pale bodies. Others have green or brown skin with white or yellow spots. Yellow stingrays are sometimes called maid stingrays or yellow spotted stingrays.

HOW TO SPOT

Size: 26 inches (66 cm) long; 14 inches (35 cm) wide

Range: Western Atlantic Ocean around the Gulf of Mexico and Caribbean Sea

Habitat: Shallow waters near sandy or muddy ocean bottoms

Diet: Clams, fish, sea worms, and shrimp

FUN FACT

Yellow stingrays have been spotted arching their bodies along the ocean floor to catch prey. The prey, such as small fish, swims under the ray, thinking the ray is a cave. Then the ray ambushes the prey and eats it.

BARNDOOR SKATE *(DIPTURUS LAEVIS)*

The barndoor skate is the largest northwestern Atlantic skate. It's also one of the largest skates in the world. These skates tend to be a reddish-brown color with dark spots and blotches. They can also have light streaks on top with light-colored underbellies. Barndoor skates have diamond-shaped pectoral fins. They are considered an endangered species.

HOW TO SPOT

Size: 30 inches (75 cm) long; 4 to 7 pounds (1.8 to 3.2 kg)

Range: Atlantic coast from southeastern Canada to Florida

Habitat: The ocean floor from coastal regions to ocean ridges, moving offshore in summer and autumn and near shore in winter and spring

Diet: Crustaceans, fish, invertebrates, small sharks, and squid

IS IT A SKATE OR A RAY?

Skates and rays look a lot alike. However, rays tend to be larger than skates. Most rays also have spines on the back of their tails to fend off predators. Skates do not have such spines. Another difference between the two is how they have their young. Female skates lay eggs, while female rays give birth to live young.

BIG SKATE *(RAJA BINOCULATA)*

The big skate can have reddish-brown or gray skin with light-colored spots and dark blotches. This diamond-shaped, bottom-dwelling skate has a pointed snout. One of its signature features is dark, eye-shaped spots that appear on its pectoral fins. It is believed that these spots help confuse the skate's predators.

HOW TO SPOT

Size: Up to 8 feet (2.5 m) long; up to 200 pounds (90 kg)

Range: Eastern Pacific Ocean from Alaska to southern California

Habitat: Near the coast in bays and estuaries, and along the continental shelf. Big skates are also seen near muddy and sandy ocean bottoms.

Diet: Clams, fish, and shrimp

FUN FACT
The big skate is the largest skate in North America.

CLEARNOSE SKATE
(RAJA EGLANTERIA)

The clearnose skate has a gray or brown color with a translucent area on either side of its snout. This is how the skate got its name. Although the skate does not have a spine on its tail to injure animals, it does have thorns on its skin. These pose potential dangers to animals and even divers who come close to a clearnose skate.

HOW TO SPOT

Size: Up to 2.7 feet (0.8 m) long; up to 7.7 pounds (3.5 kg)

Range: Northwestern Atlantic Ocean from the Gulf of Maine to the Gulf of Mexico

Habitat: Estuaries and saltwater bays near the soft, sandy ocean floor

Diet: Crustaceans, fish, and squid

FUN FACT
Like many skates, the female clearnose skate lays a black egg case. It is sometimes called a mermaid's purse.

LITTLE SKATE *(LEUCORAJA ERINACEA)*

The little skate has a rounded diamond shape, a round snout, and gray or brown skin. However, these skates can change color to match their ocean-bottom surroundings. Little skates also possess electric sensory organs in their heads known as ampullae of Lorenzini. These organs emit an electric signal to help the little skate find food. Little skates are sometimes called hedgehog skates or summer skates.

HOW TO SPOT

Size: 16 to 20 inches (40 to 50 cm) long; 1 to 2 pounds (0.5 to 1 kg)

Range: Atlantic Ocean along the eastern coast of the United States

Habitat: Shallow waters along muddy or sandy ocean bottoms

Diet: Crabs, fish, shellfish, and squid

LONGNOSE SKATE *(RAJA RHINA)*

The longnose skate earned its name for its long and pointed snout. Its skin is a brownish color that allows the skate to remain camouflaged along the sandy ocean bottom. The longnose skate sometimes burrows into the sand on the ocean floor to avoid predators. This diamond-shaped skate has two spots that resemble eyes on its dorsal fin. These spots are dark with a lighter center and border. The spots likely confuse predators that might try to eat the skate.

HOW TO SPOT

Size: Up to 4.5 feet (1.4 m) long
Range: Northeastern Pacific Ocean
Habitat: Ocean bottoms from 180 to 3,280 feet (55 to 1,000 m) deep
Diet: Crustaceans, mollusks, and worms

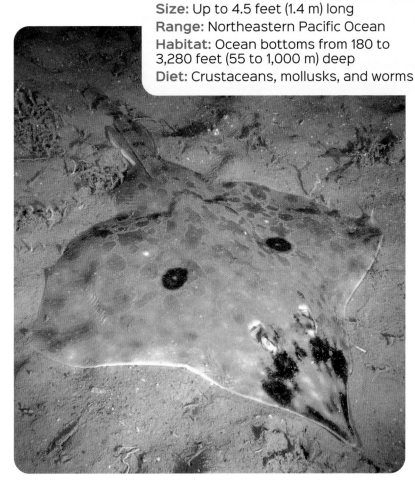

WINTER SKATE *(LEUCORAJA OCELLATA)*

Winter skates are light brown with dark spots on the top of their bodies. Underneath they have brown splotches on white skin. Their snouts are rounded, and these skates have between 72 and 110 teeth. Young skates have spines down their bodies. Winter skates are most active during the night.

FUN FACT

The winter skate can live for 19 years. Both males and females can start reproducing when they're about 11 years old.

HOW TO SPOT

Size: 3.4 feet (1 m) long; 15 pounds (7 kg)
Range: Northwestern Atlantic Ocean
Habitat: Shallow waters up to 300 feet (90 m) deep
Diet: Crustaceans, fish, mollusks, squid, and worms

ALLIGATOR GAR
(ATRACTOSTEUS SPATULA)

The alligator gar is a large freshwater fish found in North America. These fish tend to have dark, olive-brown skin. They also have a shovel-like bill that sticks out from their heads. The alligator gar is the largest of the seven species of gars that are still alive.

FUN FACT

Alligator gars are considered an ancient fish. The first gars appeared on the planet about 100 million years ago.

HOW TO SPOT

Size: 6.5 feet (2 m) long; more than 100 pounds (45 kg)

Range: Florida, Mississippi River basin, Missouri River, Texas, and Mexico

Habitat: Coastal rivers, coastal estuaries, and fresh water or brackish waters

Diet: Mainly fish, as well as dead animals, crabs, and birds

AMERICAN PADDLEFISH
(POLYODON SPATHULA)

The American paddlefish gets its name from the large paddle-like bill that extends from its snout. The fish's snout is able to detect the electric impulses of its prey. American paddlefish are considered an ancient fish species. They first appeared about 65 million years ago. These paddlefish are filter feeders. They keep their mouths open to gather food. They are sometimes called spoonbills.

FUN FACT
American paddlefish eggs can weigh about 20 pounds (9 kg).

HOW TO SPOT

Size: Up to around 7 feet (2 m) long; up to 200 pounds (90 kg)

Range: Mississippi River, Lake Huron, and parts of Canada

Habitat: Fresh water, as well as slow-moving waters deeper than 4.3 feet (1.3 m) in river basins and tributaries

Diet: Zooplankton

ATLANTIC BLUEFIN TUNA
(THUNNUS THYNNUS)

Atlantic bluefin tuna are fish that live throughout the Atlantic Ocean. These fish have metallic blue skin on the top of their bodies with silvery-white underbellies. The tuna's body is shaped like a torpedo, and its fins can retract, allowing it to easily dive 3,000 feet (900 m). Tuna must constantly swim to make sure water passes through their gills in order to oxygenate their blood.

FUN FACT

It's believed that bluefin tuna have the best eyesight of any fish.

HOW TO SPOT

Size: Up to 15 feet (4.5 m) long; up to 2,000 pounds (900 kg)

Range: From subtropical to cold regions of the eastern and western Atlantic Ocean

Habitat: Open ocean

Diet: Eels, herring, mackerel, and other fish

ATLANTIC COD *(GADUS MORHUA)*

The Atlantic cod comes in different colors, including red, olive, and yellowish green with a light-colored underbelly. This fish has a large head and a barbel that extends from its chin. Atlantic cod often swim near the ocean floor. Some of their predators include sharks, seals, and humans.

HOW TO SPOT

Size: Up to 6.5 feet (2 m) long; 88 pounds (40 kg)

Range: Northwestern Atlantic Ocean from Greenland to North Carolina

Habitat: Near ocean bottoms and in brackish waters

Diet: Crustaceans, fish, shrimp, sea cucumbers, and squid

FUN FACT

Cape Cod in Massachusetts was named for the large number of Atlantic cod found off the coast of the region.

THE COLOR-CHANGING COD

Based on what they eat, cod sometimes change color. For example, the rare brown-colored Atlantic cod gets its color from eating invertebrates such as crabs or brittle stars, which are a type of starfish. Changing color might also help cod blend into their environments.

ATLANTIC HERRING
(CLUPEA HARENGUS)

Atlantic herring have eye-catching colors, including deep blue backsides and silvery sides. These fish have small heads and a streamlined shape that make them efficient swimmers. These herring travel in large schools and feed at night. They are one of the most abundant fish in the world.

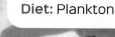

HOW TO SPOT

Size: Up to 15 inches (38 cm) long; 1.5 pounds (0.7 kg)

Range: Western Atlantic Ocean from southeastern Canada to Virginia

Habitat: Coastal waters and continental shelf waters

Diet: Plankton

BANGGAI CARDINALFISH
(PTERAPOGON KAUDERNI)

Banggai cardinalfish have small, silvery bodies with black stripes and silver spots. Their fins are long compared to their bodies, and their tail fins are forked. The pattern of spots is like a fingerprint for the fish. Each fish has a different pattern. Banggai cardinalfish can live in the wild for around three years.

HOW TO SPOT

Size: 3 inches (8 cm) long

Range: Banggai Islands of Indonesia

Habitat: Calm, shallow waters around beds of seagrass or coral reefs

Diet: Plankton and small crustaceans called copepods

FUN FACT

Banggai cardinalfish have become popular aquarium pets and are bred in captivity. However, they are sometimes taken from their natural habitat and sold. In their natural environment, they are threatened because of habitat destruction.

BELUGA STURGEON *(HUSO HUSO)*

The beluga sturgeon is anadromous, which means the fish lives in oceans but moves to freshwater rivers each year to breed. These fish have gray-black or deep blue skin with lighter underbellies. A beluga sturgeon has a long snout with barbels that extend from the bottom.

HOW TO SPOT

Size: Up to 24 feet (7.3 m) long; up to 3,500 pounds (1,600 kg)

Range: Black Sea, Caspian Sea, and Sea of Azov in Europe

Habitat: Fresh water and estuaries

Diet: Fish such as flounder, gobies, and herring

FUN FACT
Since a beluga sturgeon is large, it doesn't have a known natural predator other than people.

LARGEST AND OLDEST

The beluga sturgeon is the largest sturgeon in the world. It is also one of the largest fish in general. In addition, it can live to be one of the oldest fish. Some beluga sturgeons can live to be 100 years old.

BLACK SEA BASS
(CENTROPRISTIS STRIATA)

The black sea bass is born light gray and changes color as it ages. As an adult, the fish is gray, brown, or black. It is paler on its underbelly. This fish has a white or pale blue stripe that runs along its back. It also has dark spots on its fins. Light spots or crossbars can be seen on the sides of the fish. The black sea bass has a pointy snout and big head. Its eyes are high on top of its head.

FUN FACT

Almost all black bass start as females. Scientists have noticed that some, if not all, female bass turn into males between the ages of two and five years old.

HOW TO SPOT

Size: 12 inches (30 cm) long; up to 9.5 pounds (4.3 kg)

Range: Western Atlantic Ocean from Cape Cod to the Gulf of Mexico

Habitat: Ocean bottoms near rocky areas around buoys, pilings, and shipwrecks

Diet: Clams, crabs, small fish, shrimp, and worms

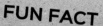

BLACK SEA BASS MATING COLOR

During mating season, the black sea bass's color changes again. Males tend to have fluorescent colors form around their eyes and the back of their necks. Females lighten in color to a blue gray or light brown.

BLACK SEA DEVIL
(MELANOCETUS JOHNSONII)

The black sea devil is one of more than 200 species of anglerfish that live in the depths of the ocean. This anglerfish is dark gray or brown. It has a large mouth that remains open, with sharp, almost clear teeth. Female anglerfish have a growth with a glowing light called a lure. It sticks out from their foreheads. This lure glows and helps the fish attract prey.

FUN FACT

The black sea devil has a lot of different names. It's also known as the humpback anglerfish, Johnson's black anglerfish, or the deep sea anglerfish.

HOW TO SPOT

Size: Up to 3.5 inches (9 cm) long

Range: Deep ocean waters off of California

Habitat: Bottom of the ocean floor at depths of up to 1,900 feet (580 m)

Diet: Invertebrates or small fish

BLOBFISH *(PSYCHROLUTES MARCIDUS)*

The blobfish is an unusual-looking fish. These fish live in the deep ocean where pressure is intense. They have soft bones. As a result, these fish don't swim but rather float above the ocean floor. If the fish are caught in fishing nets, the decrease in pressure changes their appearances. They actually look a little bit like blobs.

HOW TO SPOT

Size: Up to 12 inches (30 cm) long; up to 20 pounds (9 kg)

Range: Pacific Ocean in the deep waters off Australia and New Zealand

Habitat: Ocean floors at depths of 2,000 feet (600 m) or more

Diet: Sea urchins, sea crabs, mollusks, shellfish, and shrimp

FUN FACT
In 2013, the blobfish was voted the "World's Ugliest Animal" by the Ugly Animal Preservation Society.

BLUE MARLIN *(MAKAIRA NIGRICANS)*

Blue marlins are found throughout the world's oceans. They have silver skin with cobalt-blue backs. Blue marlins also have long bills that stick out from their upper jaws. A marlin uses this bill to stun its prey before eating it. The blue marlin is one of the fastest-growing fish. It is also one of the strongest and quickest in the ocean. There is some debate about whether Atlantic blue marlins and Pacific and Indian Ocean blue marlins are the same species.

FUN FACT

Blue marlins are a popular game fish—many people like to hunt them. Because of this, scientists see blue marlins as a vulnerable species.

HOW TO SPOT

Size: More than 16 feet (5 m) long; 1,800 pounds (820 kg)

Range: Tropical and temperate regions of the Atlantic, Indian, and Pacific Oceans

Habitat: Open oceans

Diet: Large fish such as mackerel and tuna, as well as squid

BOESEMAN'S RAINBOWFISH
(MELANOTAENIA BOESEMANI)

Boeseman's rainbowfish is native to Southeast Asia. The fish has a spectacular array of colors. The front of the fish is purple or deep blue, while the back of the fish is often orange. Stripes of green and black mark where the two colors blend. Their amazing colors have made the fish popular with home aquarium owners. The fish are considered endangered in their natural habitat.

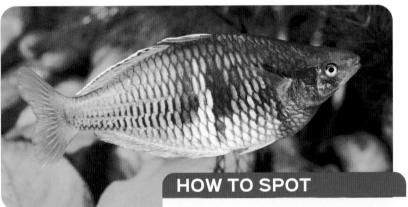

HOW TO SPOT

Size: Up to 4.5 inches (11 cm) long
Range: Western New Guinea
Habitat: Shallow, swampy areas with lots of vegetation
Diet: Algae, insects, plants, and small crustaceans

BOWFIN *(AMIA CALVA)*

The bowfin is a freshwater fish in North America. The bowfin has mottled green-and-brown skin, and its dorsal fin spreads down a long portion of its body. Males are easy to identify because they have black spots circled with orange on their tails. These fish have a variety of names, including cottonfish, dogfish, and mudfish.

FUN FACT

Bowfins are often considered primitive fish. That's because they have some qualities of fish that lived millions of years ago. For example, like some of their ancient fish ancestors, the bowfin can breathe both water and air.

HOW TO SPOT

Size: 30 inches (75 cm) long; 8.5 pounds (4 kg)

Range: North American waterways from the Saint Lawrence River and Great Lakes to eastern, southern, and western US rivers

Habitat: Clear water regions, curved lakes known as oxbow lakes, and waters that are still, without a current, called backwater regions

Diet: Fish

CLOWN ANEMONEFISH
(AMPHIPRION OCELLARIS)

Clown anemonefish are small, with orange bodies and three vertical white bars. They're sometimes confused with the orange clownfish. The major difference between the two is that orange clownfish have very obvious black border lines between their white and orange sections. Clown anemonefish have thin black lines. Sometimes, the black lines on a clown anemonefish are so thin that they are hard to make out. Depending on where the fish lives, clown anemonefish can also have white bands on a black body.

HOW TO SPOT

Size: 4 inches (10 cm) long

Range: Shallow waters of the Indian Ocean, the western Pacific Ocean, and the Red Sea

Habitat: Sea anemones

Diet: Algae, as well as drifting remains of food that sea anemones leave behind

FUN FACT

Clown anemonefish live in groups, and the biggest fish is a female. However, if the female dies, the biggest male in the group will turn into a female.

COMMON ANGELFISH
(PTEROPHYLLUM SCALARE)

Common angelfish are disk-shaped fish found in South America. In the wild, they have silver skin with black, vertical stripes. They form schools when they are young but pair off with mates as they become older. They have become a popular fish for home aquariums. As a result, they have been bred in a wide variety of colors.

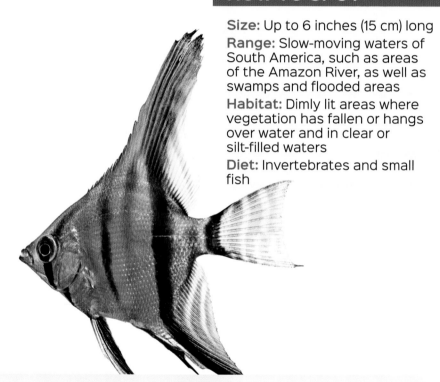

HOW TO SPOT

Size: Up to 6 inches (15 cm) long

Range: Slow-moving waters of South America, such as areas of the Amazon River, as well as swamps and flooded areas

Habitat: Dimly lit areas where vegetation has fallen or hangs over water and in clear or silt-filled waters

Diet: Invertebrates and small fish

HOME AQUARIUMS

Angelfish, goldfish, and guppies are common choices for home aquariums. But different types of fish have different needs. For instance, some need fresh water, while others need salt water. Also, some fish may not get along with other fish. In addition, there are certain fish that prefer to stay alone.

COMMON CARP *(CYPRINUS CARPIO)*

The common carp is a freshwater fish originally from Europe and Asia. Today, it is spread throughout North America. Carp have brassy-green skin with yellow underbellies. They have slimy skin, large scales, and long bodies. Each carp also has two barbels attached to its upper jaw.

HOW TO SPOT

Size: 12 to 24 inches (30 to 60 cm) long; 1 to 9 pounds (0.4 to 4 kg)

Range: Every continent except Antarctica

Habitat: Wide variety of waters, including streams and reservoirs

Diet: Plants and marine life such as crayfish, leeches, and shrimp

FUN FACT
Some scientists estimate that a female carp can lay around 300,000 eggs each breeding season.

INVASION OF THE COMMON CARP

The common carp was originally from Europe and Asia. But it probably came to North America in the 1800s. It was introduced as a game fish in the United States. Now, carp can be found in almost every US state.

COMMON FANGTOOTH
(ANOPLOGASTER CORNUTA)

The common fangtooth is solid black, brown, or gray. These fish spend much of the day in the depths of the ocean. But they have been known to come near the surface at night to feed. Because these fish have poor eyesight, they cannot get too close to the surface. Otherwise they risk becoming a meal for other fish. The fish gets its name from its large head filled with many sharp teeth. Some of these teeth resemble fangs.

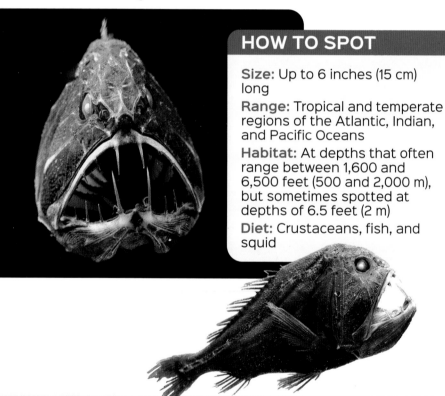

HOW TO SPOT

Size: Up to 6 inches (15 cm) long

Range: Tropical and temperate regions of the Atlantic, Indian, and Pacific Oceans

Habitat: At depths that often range between 1,600 and 6,500 feet (500 and 2,000 m), but sometimes spotted at depths of 6.5 feet (2 m)

Diet: Crustaceans, fish, and squid

SEEING IN THE DEEP SEA

There is little natural light in the deep ocean, so fish like the fangtooth have to rely on other means to see. Scientists believe that the fangtooth literally bumps into its surroundings to determine what is near it.

COMMON GOLDFISH
(CARASSIUS AURATUS)

The common goldfish is a freshwater fish that is originally from Asia. The fish are typically a yellow-gold color. However, the first common goldfish were gray or silver. People began keeping the fish as pets in China more than 1,000 years ago. Over time, breeding has caused this fish to change its natural color. Common goldfish can be black, brown, red, and even white.

FUN FACT
Goldfish have teeth in their throats.

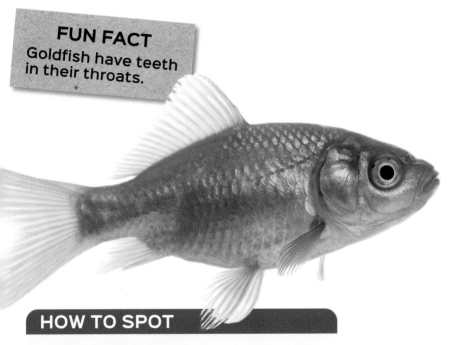

HOW TO SPOT

Size: 4.7 to 8.6 inches (12 to 22 cm) long; up to 6.6 pounds (3 kg)

Range: Native to freshwater bodies of eastern Asia but are now found in freshwater aquariums throughout the world

Habitat: Bodies of fresh water, including lakes, ponds, reservoirs, rivers, and streams

Diet: Insects, plankton, plants, and small crustaceans

COMMON GUPPY
(POECILIA RETICULATA)

The common guppy is a freshwater fish found in South America. These fish come in a variety of colors, such as blue, green, red, and yellow. The male guppies tend to be brighter and more colorful than the female guppies. Many different species of guppies have been bred for aquariums over time.

FUN FACT

Common guppies eat mosquito eggs and mosquito larvae. These fish have been used to control mosquito populations in Asia.

HOW TO SPOT

Size: Up to 2.4 inches (6 cm) long

Range: Caribbean islands and South America

Habitat: Brackish waters, clear tropical waters, and slightly salty waters

Diet: Insect larvae, shellfish, and worms

COMMON PONYFISH
(LEIOGNATHUS EQUULUS)

The common ponyfish is silver with faded bars on its back and yellow back fins. Some ponyfish have dark spots on their foreheads. This fish has a humped back, and it's able to extend its mouth downward into a tube. This allows the ponyfish to eat.

HOW TO SPOT

Size: Up to 9.8 inches (25 cm) long

Range: Indian and western Pacific Oceans from the east coast of Africa to Fiji

Habitat: Coastal waters, estuaries, river mouths, and occasionally freshwater rivers

Diet: Crustaceans, small fish, and worms

CUTLASS FISH *(TRICHIURUS LEPTURUS)*

The cutlass fish is also known as the largehead hairtail. It lives in tropical and temperate oceans throughout the world. Cutlass fish are long with silvery, almost metallic-colored skin. They have no scales and look like eels. They get their name because their bodies taper down from the head to a pointed tail. This makes them look like swords called cutlasses.

HOW TO SPOT

Size: Around 1.6 to 3.3 feet (0.5 to 1 m) long; up to 11 pounds (5 kg)

Range: Atlantic, Indian, and western Pacific Oceans

Habitat: Coastal waters and near muddy coastal ocean bottoms

Diet: Small fish and shrimp

Some birds will prey on cutlass fish.

DWARF GOURAMI
(TRICHOGASTER LALIUS)

The freshwater dwarf gourami comes in a variety of colors, including blue, red, and yellow. The fish have small bodies with large, round fins. A dwarf gourami has gills, but it also has a special organ called a labyrinth organ. It is similar to a lung and allows the fish to breathe at the surface of the water, as well as in the water.

HOW TO SPOT

Size: Up to 4.5 inches (11.4 cm) long

Range: Waterways of South Asia

Habitat: Slow-moving rivers, lakes, marshes, ponds, rice paddies, and swamps

Diet: Algae and invertebrates

FUN FACT

The dwarf gourami is a bubble nester. This means its eggs will float into a nest made of bubbles, or the male places the female's eggs into a bubble nest at the surface of the water. The fish will hatch in the bubble nest and remain there for up to three days.

EASTERN MUDMINNOW
(UMBRA PYGMAEA)

The eastern mudminnow is one of three mudminnow species that live in some freshwater areas in the United States. The eastern mudminnow is a yellowish-green color. It has ten to 12 horizontal stripes on its sides. The fish gets its name for its ability to burrow into the mud for survival during periods of high temperatures or drought.

HOW TO SPOT

Size: Up to 7 inches (18 cm) long
Range: Eastern United States
Habitat: Slow-moving bodies of water, such as backwaters and ponds
Diet: Crustaceans, insect larvae, mollusks, and worms

EUROPEAN FLOUNDER
(PLATICHTHYS FLESUS)

The European flounder is a flat fish with an oval shape and sandy-colored skin. The fish's skin includes patterns that allow the flounder to stay hidden on the sandy ocean bottom. A flounder is born with two eyes, one on each side of its head. One eye migrates to the other side of its body as the flounder ages. Generally, the flounder will end up with two eyes on one side of its body.

FUN FACT
The European flounder is also known as the North Atlantic flounder, the white fluke, or simply the flounder.

HOW TO SPOT

Size: 20 inches (50 cm) long; 6 pounds (2.7 kg)

Range: Northeastern Atlantic Ocean, Black Sea, and Mediterranean Sea

Habitat: Sandy ocean bottoms and seabeds

Diet: Crayfish, mussels, and starfish

ELECTRIC EEL
(ELECTROPHORUS ELECTRICUS)

The electric eel is long and slender with brown or dark-gray skin and a yellowish-orange underbelly. The eel has three organs that help produce an electric charge. These organs take up most of the eel's body. The eel uses these electric organs to hunt prey and stun predators.

FUN FACT

Electric eels need air to breathe, so they come to the water's surface from time to time to get oxygen.

HOW TO SPOT

Size: 6 to 8 feet (1.8 to 2.4 m) long; 44 pounds (20 kg)

Range: Northeastern areas of South America

Habitat: Murky river bottoms

Diet: Amphibians, fish, and small birds

GREEN MORAY EEL
(GYMNOTHORAX FUNEBRIS)

The green moray eel is coated with a thick mucus, which gives the eel its green appearance. However, the eel is really a dark grayish brown. The mucus on the eel helps protect it from disease and parasites. The green moray eel is long with small back fins. It is one of the largest types of morays, and its bite can be dangerous to divers.

HOW TO SPOT

Size: 6 feet (1.8 m) long; up to 65 pounds (29 kg)

Range: Eastern Atlantic Ocean from New Jersey to Bermuda and from the Gulf of Mexico to Brazil

Habitat: Protected spaces such as coral reefs, mangrove forests, and rock pilings

Diet: Crustaceans, octopuses, and small fish

FUN FACT

A green moray eel will often tie its body into a knot when hunting an octopus. This helps prevent the octopus from gripping the fish with its tentacles.

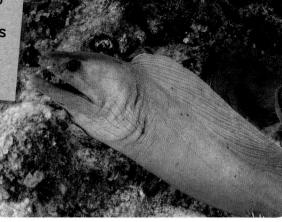

GIANT HATCHETFISH
(ARGYROPELECUS GIGAS)

The giant hatchetfish has eyes on the top of its head. Because the fish lives in darker parts of the ocean, its eyes are sensitive to light. The fish is very thin, and its body resembles the blade of a hatchet, which is how the fish got its name. The giant hatchetfish has organs called photophores along its body. A chemical reaction that occurs in these organs produces light on the fish.

HOW TO SPOT

Size: Up to 6 inches (15 cm) long

Range: Temperate and tropical oceans around the world

Habitat: Depths ranging from 1,300 to 4,000 feet (400 to 1,200 m)

Diet: Tiny fish and plankton

FUN FACT

Scientists believe the giant hatchetfish produces its light to hide from predators approaching from below the fish. The light helps the fish blend in with sunlight that penetrates the surface of the ocean.

GIANT SNAKEHEAD
(CHANNA MICROPELTES)

When it is young, the giant snakehead has black, orange, and red stripes. As it ages, the fish becomes blue black with white blotches on its sides and a white underbelly. These eel-like fish are fierce hunters. They have sharp teeth that can rip apart prey quickly and aggressively. The giant snakehead will form an *S* shape in the water before it is ready to strike with a mighty force.

FUN FACT
Giant snakeheads have been known to attack humans who get too close to their nests.

HOW TO SPOT

Size: Up to 3.3 feet (1 m) long; up to 44 pounds (20 kg)

Range: Freshwater bodies of South and Southeast Asia from India to Vietnam

Habitat: Vegetation in slow-moving freshwater bodies

Diet: Birds, fish, and frogs

GREAT BARRACUDA
(SPHYRAENA BARRACUDA)

The great barracuda is a long, cylinder-shaped fish. These fish are multicolored, with brown to blue tops, green and silvery sides, and white underbellies. They often have dark-shaded spots on their sides. These fish have sharp teeth.

FUN FACT
The great barracuda can swim up to 35 miles per hour (56 kmh) through the water.

HOW TO SPOT

Size: Up to 5 feet (1.5 m) long; 100 pounds (45 kg)
Range: Tropical and subtropical oceans
Habitat: Brackish waters near coral reefs, mangrove regions, and areas with seagrass
Diet: Fish such as anchovies, herring, and small tuna

GULF KILLIFISH *(FUNDULUS GRANDIS)*

The Gulf killifish has gray-green skin on its sides, with a pale or yellow underbelly. These fish often have spots, marks, and patterns on their skin. They have blunt snouts and lower jaws that stick out above their upper jaws. These fish are one of the largest killifish and are sometimes called chubs or bull minnows.

HOW TO SPOT

Size: Up to 10 inches (25 cm) long

Range: Coastal waters off of Cuba, Florida, and Mexico, as well as the Gulf of Mexico

Habitat: Estuaries, marshes, oyster beds, and seagrass beds

Diet: Algae, crabs, insects, mosquito larvae, and small fish

HADDOCK
(MELANOGRAMMUS AEGLEFINUS)

The haddock has gray skin with a black mark, called a thumbprint, on either side of its body above its pectoral fin. The haddock is a fast-growing member of the cod family. Many people eat this fish. As a result, haddock have been overfished in parts of the world and are considered vulnerable. The fish are sometimes called scrod.

FUN FACT
The marks on each side of the haddock are sometimes called a St. Peter's mark or a devil's thumbprint.

HOW TO SPOT

Size: Up to 3 feet (0.9 m) long; usually between 2 and 7 pounds (0.9 and 3.2 kg)

Range: Eastern and western Atlantic Ocean, with large numbers near the Canadian and US coasts

Habitat: Near the bottom of the ocean at depths between 130 and 1,000 feet (40 and 300 m)

Diet: Mollusks, sea stars, sea worms, and urchins

KING MACKEREL
(SCOMBEROMORUS CAVALLA)

The king mackerel has silver skin with a black back. It also has spots or bars on its sides. A lateral line curves along its body instead of running straight like it does in many fish. Young king mackerel have five or six rows of bronze spots on their bodies that disappear over time. The king mackerel has a torpedo shape and a pointed snout. They travel in schools.

HOW TO SPOT

Size: 1.6 to 3 feet (0.5 to 0.9 m) long; up to 99 pounds (45 kg)

Range: Western Atlantic Ocean from Massachusetts to Brazil

Habitat: Coastal waters and outer reefs

Diet: Mainly fish, crustaceans, and mollusks

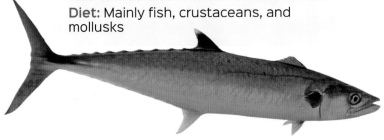

LEAFY SEA DRAGON
(PHYCODURUS EQUES)

Leafy sea dragons have green and yellow coloring with pale bands around their ribs. These fish get their name in part because of their leaf-like limbs. Their ability to look like floating seaweed helps these fish blend in with their environments. The camouflage helps them survive.

FUN FACT
Male leafy sea dragons carry fertilized eggs under their tails.

HOW TO SPOT

Size: 11.8 inches (30 cm) long
Range: Southern Australian coast
Habitat: Rocky reefs, seagrass meadows, and seaweed beds
Diet: Small crustaceans and plankton

MEKONG GIANT CATFISH
(PANGASIANODON GIGAS)

The Mekong giant catfish is a massive fish. Its back is dark to silvery gray, and its belly ranges from yellow to white. Unlike many catfish, Mekong giant catfish do not have noticeable barbels on their mouths when they are older. Young fish are born with these barbels, but they slowly shrink over time. The fish are toothless.

HOW TO SPOT

Size: Up to 10 feet (3 m) long; 650 pounds (295 kg)
Range: Mekong River in Southeast Asia
Habitat: Fresh water
Diet: Algae and plants

MEXICAN BLIND CAVE FISH
(ASTYANAX MEXICANUS)

Mexican blind cave fish are a light pink or cream color. Those that live in isolated caves are blind. This fish has a special organ on the side of its body that senses vibrations nearby. This allows the fish to swim around objects in the water and locate prey. Its skin releases mucus to help it swim quickly through the water.

FUN FACT

The Mexican blind cave fish lives in such dark environments that it lost the need for eyes. This is why the fish are blind, and some don't even have eyes.

HOW TO SPOT

Size: Up to 3.5 inches (9 cm) long

Range: From New Mexico and Texas to Guatemala

Habitat: Caverns and underground caves

Diet: Aquatic worms, small fish, snails, and plant and animal remains

MIRROR DORY *(ZENOPSIS NEBULOSA)*

The mirror dory is a flat, plate-like fish found in the Pacific Ocean. This fish has a silver-gray body with a large head and a mouth that can extend out to catch food. The mirror dory has few scales on its body and has spiny fins. One of these fins is near the mirror dory's chin.

HOW TO SPOT

Size: Usually 15.7 to 19.6 inches (40 to 50 cm) long; up to 6.6 pounds (3 kg)

Range: Coastal waters of eastern, southern, and western Australia

Habitat: Near the ocean floor around coasts or continental shelves

Diet: Crustaceans, jack mackerel, and mollusks

FUN FACT
A mirror dory can live up to 12 years, and it can start reproducing at around five years old.

MONKEY GOBY
(NEOGOBIUS FLUVIATILIS)

The monkey goby is a light brown or sandy color with dark patterns on its back. It has a black stripe that runs from its eyes to its lips. The monkey goby is different from most goby fish because its second dorsal fin gets narrower as it spreads the length of its back. The fish is also called a river goby or sand goby.

HOW TO SPOT

Size: 2.3 to 6 inches (6 to 15 cm) long

Range: Rivers in Europe, including Germany, Hungary, the Netherlands, and Turkey

Habitat: Fresh water near the shore of rivers and lakes and in estuaries and reservoirs

Diet: Crustaceans, gastropods, and fish

FUN FACT

The monkey goby is considered an invasive species. It has spread from Turkey to other waterways in Europe. This is in part because of the construction of reservoirs. The goby was also accidentally transported by getting stuck in the holes and dents along the hulls of ships that traveled to new lands.

NILE PERCH *(LATES NILOTICUS)*

The Nile perch has a long, large body that is silver and tinged with blue. There is a yellow ring around each of its dark eyes. The Nile perch also has a protruding jaw and a rounded tail. It is one of the largest freshwater fish in the world.

HOW TO SPOT

Size: 6 feet (1.8 m) long; 300 pounds (140 kg)

Range: Nile River, Lake Victoria, and other lakes and rivers of Africa

Habitat: Fresh water, particularly in tropical regions

Diet: Crustaceans, fish, insects, and mollusks

Juvenile Nile perches are silver and brown.

NORTHERN ANCHOVY
(ENGRAULIS MORDAX)

The northern anchovy has blue-green skin with a silver underbelly. An adult fish also has a silver stripe that runs along its sides. These fish have snouts that hang over their mouths. Northern anchovies live in large schools, and they are important to the ocean food chain. These fish are typically a main food eaten by other animals, such as dolphins, humpback whales, and sea lions.

FUN FACT

Anchovies are the fish that are used as a topping on anchovy pizza.

HOW TO SPOT

Size: 2.7 inches (7 cm) long

Range: Pacific Ocean from British Columbia to Baja, California, as well as the Gulf of California

Habitat: Coastal waters

Diet: Plankton

NORTHERN LAMPFISH
(STENOBRACHIUS LEUCOPSARUS)

The northern lampfish has dark skin covered with rows of photophores on its sides and belly. These produce light patterns. Both the male and female northern lampfish use the light patterns to attract mates. These fish are found in the deep ocean, where they stay to avoid encounters with predators. They move upward in the ocean at night to search for food.

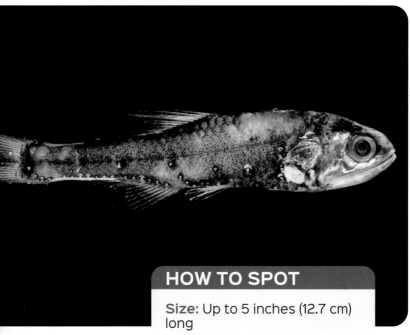

HOW TO SPOT

Size: Up to 5 inches (12.7 cm) long

Range: Eastern and western Pacific Ocean

Habitat: Deep ocean, at depths ranging from 2,300 to 3,300 feet (700 to 1,000 m)

Diet: Copepods and small fish

NORTHERN PIKE *(ESOX LUCIUS)*

A northern pike's upper sides, back, and head will range in color—between dark to light olive green. They also have white underbellies and darker blotches on both sides of their bodies. These fish have sleek bodies that allow them to move quickly and strike at prey. They are found in cold-water regions of the world and will often eat anything in their habitats.

FUN FACT

Northern pike have been known to eat animals other than fish, such as small ducklings and red squirrels. Also, a northern pike will sometimes eat other northern pike.

HOW TO SPOT

Size: Up to 4.5 feet (1.4 m) long; up to 62.5 pounds (28 kg)
Range: Freshwater bodies of the Northern Hemisphere
Habitat: Backwaters, ponds, lakes, and rivers
Diet: Amphibians, birds, fish, and invertebrates

NORTHERN PUFFER
(SPHOEROIDES MACULATUS)

The northern puffer is brown, olive, or yellow along its top and sides with a white or yellow underbelly. It has small, dark blotches or stripes on its side and spots on its back. This puffer fish is club shaped and has a small mouth that looks like a beak. The fish has a special chamber in its stomach that allows it to take in air or water. This makes the puffer fish expand in size to frighten predators.

HOW TO SPOT

Size: 8 to 10 inches (20 to 25 cm) long

Range: Western Atlantic Ocean from southeastern Canada to Florida

Habitat: Bays and estuaries at a depth of 195 feet (60 m)

Diet: Crustaceans, mollusks, and small invertebrates

FUN FACT

If a northern puffer fish is caught while it is inflated with air and thrown back in the water, it will float on the surface upside down. It will soon deflate and swim away.

NORTHERN STARGAZER
(ASTROSCOPUS GUTTATUS)

The northern stargazer has a blackish-brown body with white spots. It also has three horizontal lines on its tail. The stargazer's two eyes are on the top of its head. This fish uses its fins to burrow into the sand and waits for its prey to swim by before attacking. It also has an electric sensor on its head that it uses to stun prey.

HOW TO SPOT

Size: 8 to 22 inches (20 to 55 cm) long; up to 20 pounds (9 kg)

Range: Subtropical to cold regions of the Atlantic Ocean from New York to North Carolina

Habitat: Depths of 120 feet (36.5 m) on sandy or rocky ocean bottoms

Diet: Crustaceans, such as crabs, and small fish

OARFISH *(REGALECUS GLESNE)*

The oarfish has silvery skin with markings and wave patterns, as well as long fins that are red or pink. The fish lives deep in the ocean and has large eyes. An oarfish has no teeth and swims with its mouth open to gather food such as krill. The food is collected inside its mouth on bones that help support the fish's gills. These bones act like a rake and strain food the fish takes in. The oarfish is sometimes called a ribbonfish.

HOW TO SPOT

Size: 10 feet (3 m) long; up to 600 pounds (272 kg)

Range: Tropical and temperate oceans

Habitat: Deep, dark regions of open ocean to depths of 3,000 feet (915 m)

Diet: Krill, plankton, and small crustaceans

FUN FACT

Some oarfish have been spotted at the surface with their snake-like bodies. Such sightings may have led to sea serpent legends.

Krill are small—sometimes only 2.4 inches (6 cm) long.

83

OCEAN SUNFISH *(MOLA MOLA)*

The ocean sunfish is blue, gray, or white. It has jagged scales that make its skin rough and gritty to the touch—almost like sandpaper. These fish are also the heaviest bony fish in the world. The ocean sunfish can be seen at the surface of the ocean, basking in the sun. Because its dorsal fin sticks out, a sunfish can occasionally be mistaken for a shark.

FUN FACT

A female ocean sunfish can lay 300 million eggs. That's more than any other fish species.

HOW TO SPOT

Size: Up to 11 feet (3.3 m) long; average weight of 4,000 pounds (1,800 kg)

Range: Atlantic, Indian, and Pacific Oceans, as well as the Mediterranean and North Seas

Habitat: From the surface to typical depths of 557 feet (170 m) and within 125 miles (200 km) of the coast

Diet: Algae, fish, jellyfish, and zooplankton

OPAH *(LAMPRIS GUTTATUS)*

The opah has a unique and colorful appearance. This fish has silver skin with red fins and a red mouth. The belly of the opah contains reddish shades with white spots. A gold band circles the opah's eyes. It has a round, flat body.

HOW TO SPOT

Size: Up to 3 feet (0.9 m) long; 100 pounds (45 kg)

Range: Tropical and temperate oceans of the world

Habitat: Deep oceans during the day and the surface at night

Diet: Krill, squid, and small fish

SLEEPING FISH

Fish don't sleep like land mammals do. Instead of closing their eyes and lying down for the night, fish begin moving slower. Some even wedge their bodies into safe places, float without moving much, or retire to a nest. While fish might be less active at certain points of the day, they remain alert and watch for predators.

ORNATE GHOST PIPEFISH
(SOLENOSTOMUS PARADOXUS)

The ornate ghost pipefish has a very distinct look, with thin projections on its fins and body. This species can also be a variety of colors, such as black, red, yellow, or semitransparent. It also can have dots, lines, or other blotches on its body. The ornate ghost pipefish has a long snout which it uses to suck up food.

HOW TO SPOT

Size: Around 5.1 inches (13 cm); males are smaller than females

Range: Indian and Pacific Oceans

Habitat: Coastal waters near rocky or coral drop-offs

Diet: Small aquatic creatures

FUN FACT
An ornate ghost pipefish can change its color depending on its environment.

PACIFIC SEAHORSE
(HIPPOCAMPUS INGENS)

The Pacific seahorse comes in colors such as brown, gold, red, or yellow. These fish often have vertical white markings on their bodies. Their heads have bony snouts that resemble a horse's, which gave rise to the fish's name. This fish doesn't have teeth or a stomach, so it relies on its snout to eat. When the fish snaps its head back, it enables the fish to suck in its food from the surrounding water through its snout.

HOW TO SPOT

Size: 5.1 to 11 inches (13 to 28 cm) long

Range: Pacific Ocean from southern California to Peru

Habitat: Coral reefs, at depths often ranging from 10 to 60 feet (3 to 18 m)

Diet: Small crustaceans and zooplankton

FUN FACT
Male seahorses have pouches that hold seahorse eggs. The eggs hatch inside the pouch and are carried there until the male seahorse births the live young.

PETER'S ELEPHANTNOSE FISH
(GNATHONEMUS PETERSII)

The Peter's elephantnose fish is typically dark brown or black. It has a long snout that extends from its chin and helps it eat prey. These fish were once thought to be blind. But scientists have learned that they have unique eyes that can easily see objects in murky water.

HOW TO SPOT

Size: Up to 9 inches (23 cm) long
Range: African regions
Habitat: Freshwater rivers with lots of vegetation, and in muddy and dark riverbeds
Diet: Insect and mosquito larvae

PINECONE FISH
(MONOCENTRIS JAPONICA)

The pinecone fish is also known as the Japanese pineapple fish. It has bright yellow or orange skin with distinctive black outlines. These outlines make the fish resemble a pinecone or pineapple. These fish have small tails and a black lower jaw.

HOW TO SPOT

Size: 4 to 5 inches (10 to 12.7 cm) long

Range: Indian and Pacific Oceans

Habitat: Ocean areas with caves, coral reefs, or rocks

Diet: Shrimp and small fish

FUN FACT

The pinecone fish has an organ on the sides of its head that helps it make light. This light might be used to communicate with other fish or to catch prey.

POTATO GROUPER
(EPINEPHELUS TUKULA)

The potato grouper has gray-brown skin with brown marks. These marks look like potatoes. The potato grouper is one of the largest fish to live in coral reefs. This fish often waits in the coral, ready to pounce on prey that swims by. The potato grouper has a large mouth that it uses to swallow its food whole. This fish is sometimes called the potato cod, potato rockcod, or potato bass.

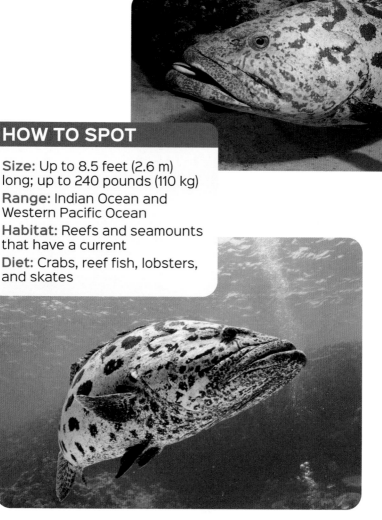

HOW TO SPOT

Size: Up to 8.5 feet (2.6 m) long; up to 240 pounds (110 kg)

Range: Indian Ocean and Western Pacific Ocean

Habitat: Reefs and seamounts that have a current

Diet: Crabs, reef fish, lobsters, and skates

QUEEN PARROTFISH
(SCARUS VETULA)

The queen parrotfish comes in many bright colors, such as blue, green, and orange. These fish have markings above their eyes that might look like a crown. The parrotfish has about 1,000 teeth found in 15 rows that appear fused together to form two biting plates. These plates are located outside of the parrotfish's mouth. This makes it looks like the fish has a beak—giving it a parrot-like appearance. At night, queen parrotfish cover themselves in mucus. They most likely do this to mask their scents from predators.

FUN FACT
All queen parrotfish are females when they hatch. As they grow, the largest turn into male parrotfish.

HOW TO SPOT

Size: 1 to 4 feet (0.3 to 1.2 m) long

Range: Tropical reefs throughout the world

Habitat: Coral reefs

Diet: Algae

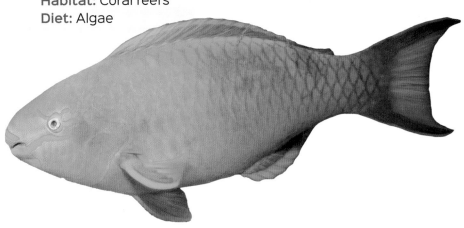

RAINBOW TROUT
(ONCORHYNCHUS MYKISS)

The rainbow trout comes in a variety of spectacular colors. The color depends on the age, habitat, and mating condition of each fish. They are typically bluish green or yellowish green with white bellies, and pink streaks are found along their sides. They have a torpedo shape and often have small black spots along their backs and dorsal fins.

HOW TO SPOT

Size: 20 to 30 inches (50 to 76 cm) long
Range: North America west of the Rocky Mountains
Habitat: Fresh water; cool, clear lakes; rivers; and streams
Diet: Crustaceans, insects, and small fish

FUN FACT

Some rainbow trout will leave freshwater rivers to live in the ocean. These fish are known as steelheads and their skin becomes silver.

RED-BELLIED PIRANHA
(PYGOCENTRUS NATTERERI)

Red-bellied piranhas are gray with bright silver scales. The belly, cheeks, and chin of this fish are red, which give the fish its name. These piranhas have strong jaws and razor-sharp teeth. The teeth on the bottom set of the fish's mouth lock with the upper set. They can chomp down with enough force to bite through bone. They often travel in schools of up to 20 fish.

HOW TO SPOT

Size: A little over 12 inches (30 cm) long; 4 pounds (1.8 kg)

Range: South America

Habitat: Freshwater rivers, especially the lower Amazon River basin

Diet: Algae, insects, invertebrates, plants, small fish, and the ends of fish tails

RED-LIPPED BATFISH
(OGCOCEPHALUS DARWINI)

The red-lipped batfish has light brown and gray skin on its back with a white underbelly. The batfish usually has dark brown dots running from its snout to its tail. The fish has bright red coloring around its mouth. It has a back fin that is used as a lure to attract prey. The red-lipped batfish is unusual because it often uses its fins like legs to walk or rest on the ocean floor.

HOW TO SPOT

Size: Up to 7.9 inches (20 cm) long

Range: Around the Galápagos Islands in the Pacific Ocean

Habitat: The sandy ocean floor near reefs

Diet: Invertebrates, small crustaceans, small fish, and worms

FUN FACT
Scientists believe the red-lipped batfish's brightly colored mouth could be a way for the fish to attract a mate.

RED PORGY *(PAGRUS PAGRUS)*

The red porgy has a pinkish-red color with blue spots that cover its body. It also has yellow coloring around its eyes and along its snout. These fish have a large, round appearance caused by their humped backs. The fish often swim in schools, traveling together in search of food.

HOW TO SPOT

Size: Up to 36 inches (91 cm) long; up to 20 pounds (9 kg)
Range: Atlantic Ocean
Habitat: Coastal regions along rocky or sandy ocean bottoms
Diet: Crustaceans, mollusks, and small fish

GOING TO SCHOOL

Up to 80 percent of all young fish travel in a large group called a school. Traveling in schools helps keep fish somewhat safe from predators. A school of small fish might look like a large creature. This makes a predator less likely to attack. In addition, schools often have lookout fish. These fish keep an eye on predators that might attack. They might move suddenly in response to a predator. As a result, the whole school might suddenly change direction.

REEDFISH *(ERPETOICHTHYS CALABARICUS)*

Reedfish are an eel-like fish native to West Africa. These fish are generally green with yellow underbellies. Reedfish take in oxygen through both their gills and skin. They also have lungs. These fish come to the surface to breathe air several times a day. Reedfish are sometimes called ropefish or snake fish.

FUN FACT

Because it has lungs, the reedfish can stay alive for about eight hours without being in water, though its skin needs to be moist during this time.

HOW TO SPOT

Size: Up to 11.8 inches (30 cm) long
Range: Freshwater rivers of Benin, Cameroon, and Nigeria
Habitat: Floodplains and streams in tall grasses and reeds
Diet: Crustaceans, insect larvae, small fish, and worms

SAILFIN FLYING FISH
(PAREXOCOETUS BRACHYPTERUS)

The sailfin flying fish has dark green or blue skin on the top of its body with silvery skin on the rest. All fins except for the tail and pelvic fins are clear. The tail fin has some red coloring. These fish also have large, black spots on their dorsal fins.

HOW TO SPOT

Size: Up to 7.8 inches (20 cm) long

Range: Equatorial waters of the Atlantic, Indian, and Pacific Oceans

Habitat: Ocean depths between 0 and 65 feet (0 and 20 m)

Diet: Fish eggs and plankton

FUN FACT
The sailfin is able to fly by beating its tail up to 70 times per second. This speed allows the fish to propel itself out of the water.

THE FLYING FISH NAME
The sailfin flying fish gets its name for its enlarged pectoral fins that look almost like wings. These fins help the fish glide in the air when it jumps out of the water to escape predators.

SILVER SALMON
(ONCORHYNCHUS KISUTCH)

Silver salmon are also known as coho salmon. These fish have silver sides, and they have black spots on their backs. For the first year of their lives, silver salmon will live in fresh water. Then, they swim into the ocean. These fish will go back into freshwater streams to breed. At this time, males will have dark red coloring on their sides. They will also develop hooked jaws.

HOW TO SPOT

Size: 2 to 2.5 feet (0.6 to 0.76 m) long; 8 to 12 pounds (3.6 to 5.4 kg)

Range: Large rivers and small coastal streams, as well as the northern Pacific Ocean

Habitat: Slow-moving freshwater and oceans closer to shore

Diet: Fish and insects

SOCKEYE SALMON
(ONCORHYNCHUS NERKA)

The sockeye salmon can survive in both fresh water and salt water. When heading into the ocean, these fish have silver bodies with blue-green backs and white underbellies. However, when the salmon return to fresh water in order to reproduce, their colors change. Their heads turn green, and their bodies become red. Their jaws also get longer and their noses develop a hook shape.

FUN FACT
Sockeye salmon produce orange meat. They get this color from the orange krill they eat when they live in the ocean.

HOW TO SPOT

Size: 1.5 to 2.5 feet (0.45 to 0.76 m) long; 2.2 to 8.8 pounds (1 to 4 kg)

Range: Pacific Ocean from Alaska to northern California

Habitat: Freshwater lakes, rivers, streams, as well as the ocean

Diet: When they're in fresh water, the salmon will eat amphipods, insects, and zooplankton. When they're in salt water, the fish will eat zooplankton, fish, squid, and krill.

CALIFORNIA SCORPION FISH
(SCORPAENA GUTTATA)

The California scorpion fish can be reddish brown, white, tan, and lavender, and it has dark spots and a large head. Adult California scorpion fish will have large, organized spots on their tails, as well as on their dorsal, anal, and pectoral fins. Juveniles don't have this type of spotting. California scorpion fish will stay at the bottom of the ocean during the day and hunt at night. They have poisonous spines.

HOW TO SPOT

Size: Can be around 18 inches (45 cm) long

Range: Eastern Pacific Ocean

Habitat: Rocky sections in bays and in caves

Diet: Small fish and crabs, octopuses, and shrimp

FUN FACT
The scorpion fish's appearance helps it stay camouflaged in its rocky and coral surroundings.

SPOTTED SCORPION FISH
(SCORPAENA PLUMIERI)

The spotted scorpion fish comes in different colors, such as shades of black and brown on a light background, or red or orange coloring. The colors act as camouflage to help the fish hide in its ocean-bottom surroundings. Skin flaps or feather-like fins cover the scorpion fish to add to this camouflage. The fish also has poisonous spines on its dorsal fins. These spines contain venom that is used to stun prey.

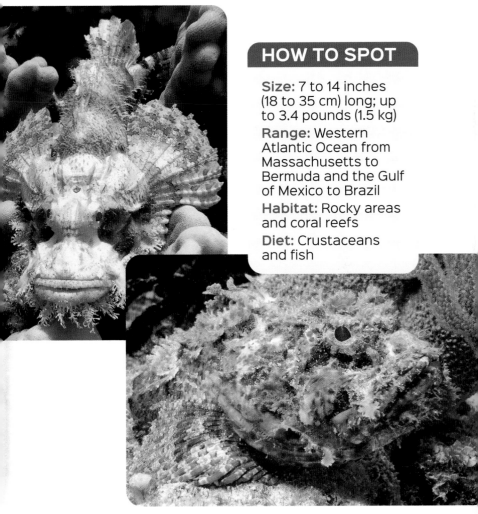

HOW TO SPOT

Size: 7 to 14 inches (18 to 35 cm) long; up to 3.4 pounds (1.5 kg)

Range: Western Atlantic Ocean from Massachusetts to Bermuda and the Gulf of Mexico to Brazil

Habitat: Rocky areas and coral reefs

Diet: Crustaceans and fish

SWORDFISH *(XIPHIAS GLADIUS)*

Swordfish live in tropical, temperate, and some cold-water ocean regions. The fish have dark brown or black coloring at the top of their bodies. The skin color lightens toward their bellies. Adult members of the species have no scales or teeth. Instead, they rely on the large, flat bills that extend from their snouts. They most likely slash their prey, and small prey can be eaten whole.

FUN FACT

A swordfish has a special organ next to its eyes to help keep its eyes and brain warm.

HOW TO SPOT

Size: 47 to 75 inches (120 to 190 cm) long; more than 300 pounds (140 kg)

Range: Atlantic, Indian, and Pacific Oceans

Habitat: Depths ranging from 650 to 1,970 feet (200 to 600 m)

Diet: Crustaceans, fish, and squid

YELLOW GOATFISH
(MULLOIDICHTHYS MARTINICUS)

The yellow goatfish generally has a silver body with yellow fins and a long, yellow line that extends along its body. These fish forage for food along reefs and ocean bottoms. They have two barbels extending from their chins. This helps them find food.

HOW TO SPOT

Size: 6 to 12 inches (15 to 30 cm) long

Range: Atlantic Ocean

Habitat: Sandy regions of lagoons and seaward reefs

Diet: Brittle stars, crustaceans, small fish, and sea worms

FUN FACT
The yellow goatfish can change color to blend in with its environment.

INDONESIAN COELACANTH
(LATIMERIA MENADOENSIS)

The Indonesia coelacanth was first discovered in 1998. It is a species of primitive fish, and it resembles ancient fish that lived during the prehistoric age of dinosaurs. These fish are a brownish color with spots.

HOW TO SPOT

Size: Up to 4.6 feet (1.4 m) long

Range: Pacific Ocean off the coast of Indonesia

Habitat: Often in caves, near rocky slopes of volcanic islands, at depths of 492 to 656 feet (150 to 200 m)

Diet: Octopuses, small fish, and squid

FUN FACT
Coelacanths are often called fossil fish because they lived when dinosaurs did. The first coelacanths appeared 360 million years ago.

WEST INDIAN OCEAN COELACANTH *(LATIMERIA CHALUMNAE)*

West Indian Ocean coelacanths have dark-colored skin with light underbellies and spots on their sides. The West Indian Ocean coelacanth and its closest relative, the Indonesian coelacanth, are the only vertebrates that have a jointed skull. This helps their mouths open wider for food. They are nocturnal hunters, spending the day in underwater caves.

HOW TO SPOT

Size: More than 6.5 feet (2 m) long
Range: Western Indian Ocean
Habitat: Twilight zone region of the ocean, at depths between 500 and 800 feet (152 and 243 m) in lava caves
Diet: Invertebrates and small fish

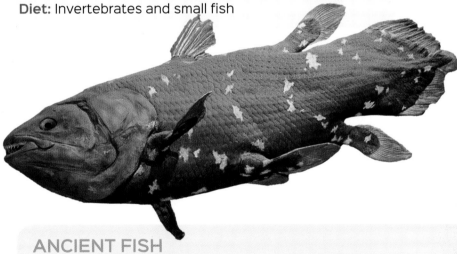

ANCIENT FISH

For years, coelacanths were thought to be an ancient fish that died with the dinosaurs. Based on fossil information, it had looked like the last coelacanth disappeared 65 million years ago, around the time dinosaurs went extinct. However, in 1938 a West Indian Ocean coelacanth was caught alive off the coast of South Africa. It did not resemble any of the previously known coelacanths, so it was believed to be part of a new coelacanth species. Since that time, only one other species of coelacanths has been found alive—this time in Indonesia.

AUSTRALIAN LUNGFISH
(NEOCERATODUS FORSTERI)

The Australian lungfish has brown or olive-green skin with a white underbelly and blotches on its sides. The Australian species has one lung. It can use this lung during dry periods to breathe air instead of using its gills. When this happens, the lungfish tends to breathe one to two times per hour. These fish have large scales that cover their bodies.

HOW TO SPOT

Size: Up to 4.9 feet (1.5 m) long

Range: Queensland region of Australia, along the Burnett and Saint Mary's River

Habitat: Fresh water, especially still or slow-moving waters

Diet: Earthworms, frogs, small fish, snails, and tadpoles

RAY-FINNED FISH VS. LOBE-FINNED FISH

The two classes of bony fish are ray-finned fish and lobe-finned fish. While both fish have bones, there is a slight difference in their structures. A ray-finned fish has fins that are made of webbed rays. Their pectoral fins have a few bones. Lobe-finned fish have muscular fins. There's one bone that joins the fins to the body.

SOUTH AMERICAN LUNGFISH
(LEPIDOSIREN PARADOXA)

The South American lungfish has gray or black skin. When these fish are young, they have bright yellow spots. South American lungfish have slender bodies and long pectoral and pelvic fins. Because their eyesight is poor, the fish use these fins to touch and sense their surroundings. During times of drought, when water is scarce, they can burrow in the mud so they do not become too dry.

FUN FACT

South American lungfish are born with gills to breathe in the water, but the gills begin to stop functioning after seven weeks. Then, the fish must start breathing air at the water's surface.

HOW TO SPOT

Size: Up to 4 feet (1.2 m) long

Range: South America, mainly in the region of the Amazon River basin

Habitat: Fresh water, lakes, still waters with little current, and swamps

Diet: Algae, fish, insects, plants, shrimp, and snails

GLOSSARY

barb
A sharp point on the end of a ray's or skate's tail.

barbel
A fleshy growth on the mouth of a fish.

camouflage
Patterns or colors used for hiding.

cartilage
Firm, flexible tissue that forms the skeleton of some fish.

dorsal
Relating to the upper side and back of a fish.

estuaries
Parts of a river where fresh water meets the ocean.

gastropod
An invertebrate such as a snail or slug.

invasive species
A species that is introduced to a new area and causes damage to the habitat.

kelp
Large, brown seaweed.

krill
A small, shrimp-like crustacean.

mangrove forests
Trees and shrubs that live in coastal regions near the equator where ocean water is present at high tide but not at low tide.

pectoral
Relating to the region behind a fish's head, often where fins are found.

pelvic
Relating to the fins on the bottom of a fish.

silt
Fine sand or clay that is carried by moving water.

volt
A measurement of electric force.

TO LEARN MORE

FURTHER READINGS

Harvey, Derek. *Sharks and Other Deadly Ocean Creatures*. DK Publishing, 2016.

Hogan, Zeb. *Monster Fish!* National Geographic, 2017.

Wilsdon, Christina. *Ultimate Oceanpedia*. National Geographic, 2016.

ONLINE RESOURCES

To learn more about fish, please visit **abdobooklinks.com** or scan this QR code. These links are routinely monitored and updated to provide the most current information available.

PHOTO CREDITS

Cover Photos: Howard Chen/ iStockphoto, hammerhead shark; Vlad Oskan/iStockphoto, great white shark; iStockphoto, lemon shark, Banggai cardinalfish, sturgeon, clown anemonefish, goldfish, back (queen parrotfish), back (yellow goatfish); Global Pics/ iStockphoto, manta ray; Mirko Rosenau/iStockphoto, angelfish, guppy, dwarf gourami; iStockphoto, electric eel; Shutterstock Images, Boeseman's rainbowfish; Karel Bartik/Shutterstock Images, rainbow trout; Dennis Jacobsen/ Shutterstock Images, Atlantic cod; Steven Russell Smith Ohio/Shutterstock Images, sockeye salmon; Joe Belanger/ iStockphoto, back (ornate ghost pipefish); Nantawat Chotsuwan/ Shutterstock Images, back (pineapple fish)

Interior Photos: iStockphoto, 1 (zebra shark), 1 (alligator gar), 1 (goldfish), 1 (great barracuda), 5 (sea lamprey), 5 (leafy sea dragon), 7, 9 (top), 9 (bottom), 14, 15 (top), 15 (bottom), 19 (top), 19 (bottom), 20, 21 (top), 24 (top), 26 (top), 26 (bottom), 40 (top), 42, 45 (top), 46, 51 (top), 52, 53 (top), 53 (bottom), 55, 57, 58 (bottom), 60 (bottom), 64 (top), 64 (bottom), 65 (bottom), 68 (top), 72 (top), 73, 74 (top), 74 (bottom), 80 (bottom), 83 (bottom), 84 (top), 86 (bottom), 87 (top), 87 (bottom), 89, 90 (bottom), 92 (top), 92 (bottom), 93 (top), 101 (top), 101 (bottom), 112 (great white shark), 112 (Mexican blind cave fish), 112 (sea horse); Global Pics/iStockphoto, 1 (manta ray), 28 (top), 28 (bottom); Saran Jantraurai/Shutterstock Images, 4 (American paddlefish), 41 (top); Mirko Rosenau/iStockphoto, 4 (dwarf gourami), 5 (guppy), 58 (top), 61 (top), 61 (bottom); Joe Quinn/ Shutterstock Images, 4 (queen parrotfish), 91; Andy Krakovski/ iStockphoto, 5 (northern pike), 5 (sockeye salmon), 80 (top), 99 (top); Shutterstock Images, 6, 13, 40 (bottom), 43, 44, 50 (top), 51 (bottom), 54, 88 (top), 94 (bottom), 95; Tom McHugh/Science Source, 8, 77 (top); Martin Prochazkacz/ Shutterstock Images, 10; Norbert Probst/imageBROKER/ SuperStock, 11 (top), 11 (bottom); Kelvin Aitken/VWPics/AP Images, 12; Jeffrey Rotman/Biosphoto/ Alamy, 16; JimCatlinPhotography. com/Shutterstock Images, 17 (top); Shane Gross/Shutterstock Images, 17 (bottom); Howard Chen/iStockphoto, 18; Deborah Holden/iStockphoto, 21 (bottom); Andy Murch/Blue Planet Archive, 22, 34, 35, 36, 37 (top), 37 (bottom), 38; Andy Murch/ VWPics/Newscom, 23 (top), 30

ABDOBOOKS.COM

Published by Abdo Publishing, a division of ABDO, PO Box 398166, Minneapolis, Minnesota 55439. Copyright © 2021 by Abdo Consulting Group, Inc. International copyrights reserved in all countries. No part of this book may be reproduced in any form without written permission from the publisher. Abdo Reference™ is a trademark and logo of Abdo Publishing.

Printed in the United States of America, North Mankato, Minnesota.
082020
012021

Editor: Alyssa Krekelberg
Series Designer: Colleen McLaren

Library of Congress Control Number: 2019954396
Publisher's Cataloging-in-Publication Data
Names: Forest, Christopher, author.
Title: Fish / by Christopher Forest
Description: Minneapolis, Minnesota : Abdo Publishing, 2021 | Series: Field guides for kids | Includes online resources and index.
Identifiers: ISBN 9781532193057 (lib. bdg.) | ISBN 9781098210953 (ebook)
Subjects: LCSH: Fishes--Juvenile literature. | Fishes--Behavior--Juvenile literature. | Fishes--Field guides--Juvenile literature. | Reference materials--Juvenile literature.
Classification: DDC 597--dc23